BIRMINGHAM

Images at and accessible to the Transport Treasury archive

200

Years of Train Travel
Since 1825

ISBN 978-1-917776-14-1

First published in 2025 by Transport Treasury Publishing Ltd. 16 Highworth Close, High Wycombe, HP13 7PJ
Totem Publishing, an imprint of Transport Treasury Publishing.

www.ttpublishing.co.uk

Printed in Tarxien, Malta by Gutenberg Press Ltd.

Front cover - Stanier Jubilee 45569 *Tasmania* emerges from the tunnels of Worcester Street and the bridge of Queens Drive into the Midland side of New Street station. The single tunnel at the east end of New Street was a bottleneck and in 1895 an additional tunnel was added enabling four lines from Proof House Junction. The LNWR design New Street No. 2 Signal Box was solely dedicated to the operation of the Midland side of the station from then onwards. 45569 was a Leeds Holbeck engine. *(R C Riley)*

Frontispiece - The outside of New Street Station fronting onto Stephenson Street. Queens Hotel on the right hand side. Behind the photographer is Stephenson Place leading to New Street itself. On the corner was a bank and next door (but out of shot) the Midland Hotel. It used to be home to Atkinsons Bar which was small, very cosy, and for a long time, a real ale bar. A very different view is presented now with West Midlands Metro trams running along Stephenson Street with a stop outside the entrance into the modern New Street station. *(Arthur Mace)*

Rear cover - Twin class 25's arriving at Bescot from the Walsall direction with a rake of 16t mineral wagons. 25297 was built as D7647 entering traffic in March 1966 and being withdrawn in September 1986. 25178 was new to traffic as D7528 in February 1965 and withdrawn in April 1985. Both engines would be scrapped at Vic Berry's scrapyard in 1989. *(Arthur Turner)*

INTRODUCTION

Birmingham: Britain's second city, and known as the city of a thousand trades, workshop of the world, and possessing more canal miles than Venice. It is situated in the heart of the Midlands and was the very epicentre of the industrial revolution. The first railway in the region was the Shutt End Railway linking the coal mines of Pensnett and Kingswinford with the Staffordshire and Worcestershire canal which opened in 1829, with the James Foster and John Rastrick built *Agenoria*, which today can be found at York in the Railway Museum.

Phillip Hardwick's impressive building at Curzon Street. Built in 1838 and terminus of the London and Birmingham Railway and Grand Junction railway and mirrored the Euston Arch. Alongside can be seen the first Queens Hotel built in 1846. Its life as a railway station was short as traffic growth soon expanded to more than the site could cope with and closed to passenger use in 1854. *(George Barlow)*

With the success of the Stockton and Darlington Railway in 1825 and Liverpool and Manchester Railway in 1830, the first line into the city was the Grand Junction Railway arriving in 1837 to a temporary terminus at Vauxhall with a line from the Liverpool and Manchester, through Warrington and Wolverhampton to Birmingham. Completion of the London and Birmingham in 1838 saw both lines move to the new terminus at Curzon Street. Meanwhile the Birmingham and Derby Railway and Birmingham and Gloucester Railway arrived in the 1840's being linked by the Camp Hill Line. The Great Western had designs on London to Liverpool traffic and the Birmingham to Oxford Railway sought to create a GWR line into Curzon Street, a move the LNWR tried to resist at some cost to the GWR as will be seen later.

The GWR station at what is now known as Snow Hill opened in 1852 with the connection to Wolverhampton north being completed in 1854. The capacity of Curzon Street was rapidly outstripped and in 1854 the now London and North Western Railway opened New Street, a 'Grand Central' station right in the heart of Birmingham and at a crossroads of lines from Derby, Bristol, London, Liverpool and Manchester.

The Birmingham West Suburban Railway, which opened in 1876 from Lifford on the Camp Hill line to a terminus not far from Five Ways, was extended into the new Midland Railway platforms at New Street in 1885. On the Great Western Railway, Snow Hill was rebuilt in 1871 and then again between 1906 and 1912. To increase capacity for suburban services, the GWR opened Birmingham Moor Street in 1910 and the North Warwickshire line opened in 1908, providing the GWR with a second line from Birmingham to Bristol, running as it did through Stratford upon Avon and down through Honeybourne to Cheltenham.

One thing of note on the topography of Birmingham is that both Snow Hill and New Street are constrained by the need for tunnels at either end of the station and a cause for some of the developments that followed.

Birmingham was a city building railway rolling stock. Bellis and Seeking on Broad Street were an early builder but better known are Metropolitan Cammell, builders of carriages and wagons and Birmingham Railway Carriage and Wagon (BRCW) at Smethwick who are most known for construction of Class 26,27 and 33 diesel locomotives. Last construction of railway rolling stock took place at Alstom, successors of Metropolitan Cammell at Washwood Heath with their Pendolino sets.

As you may expect Birmingham was home to a number of engine sheds with Monument Lane, Bescot, Aston, Saltley and Bournville on the LMS side and Tyseley on the Great Western. There was also a breadth of motive power as will be seen in the following pages from venerable Victorian and Edwardian engines through to the express engines of the GWR and LMS along with occasional visitors from other regions.

Birmingham was also home to a number of industrial railways with the Cadbury system at Bournville, Austin at Longbridge, Dunlop, and Nechells power station and I have tried to feature each.
The following pages track the growth and contraction of the railways of Birmingham, the closures of the 1960's, most notable being the closure of Snow Hill and the through GWR main line to Wolverhampton and then its reopening in 1987 and the line's later inclusion into the Midland Metro. The Snow Hill reopening then led to the closure of Moor Street terminus platforms, but these in turn reopened in 2010.

At the time of writing new stations on the Camp Hill line are being built with its reopening to passenger traffic planned at the end of 2025 over eighty years after they were closed. Curzon Street in a remarkable turn of events, that historic station of the Grand Junction Railway and London and Birmingham railway is to be the terminus of HS2 with massive building works under way. Even the trams which left the city in 1952 have returned with the Midland Metro running from Wolverhampton through to Edgbaston with extensions to Dudley and Eastside being built.

Birmingham can also be considered the birthplace of the preservation movement with the founding of the Talyllyn Railway Preservation Society and the inaugural meeting at now long lost Imperial Hotel. Patrick Whitehouse and Pat Garland then went on to acquire firstly Small Prairie 4555, then 64xx 6435 and 14xx 1420 for the Dart Valley Railway, but before they went to Devon found a home at Tyseley shed. Also purchased was 7029 *Clun Castle* which also made its preservation home at Tyseley and never left. Tyseley is now home to Vintage Trains which became a main line train operating company and runs regular steam excursions as well as the Shakespeare Express to Stratford-upon-Avon on summer Sundays. It is also home to Tyseley Locomotive works who undertake the overhaul and construction of engines playing an integral part in a number of new builds including completion of 6880 *Betton Grange* in 2024.

My thanks to the Transport Treasury for their invitation to compile this book on Birmingham and marking Rail 200. A big thank you to Andrew Royle for pulling out the images and then finding more as I thought of locations to include. My thanks to my wife Jane, and sons Ewan and Dylan for their support and encouragement. Reference was made to the excellent BR Database for loco allocations and Six Bells Junction for railtour details and acknowledgement to the excellent Warwickshire Railways website. Too many books to list were consulted and my thanks to those authors for their work in allowing me to tell the story of railways in Birmingham.

With just a few days to go to the declaration of World War 2 the view at the north end of Birmingham Snow Hill sees GWR Saint 2903 Lady of Lyon on a short set of coaches. The absence of lamps on the engine suggests possibly a shunting movement into the carriage sidings in the background. Lead vehicles are early Collett designed examples. On the left is Snow Hill North signal box which with limited space was built on a steel gantry and all points and signals electrically operated. The line drops down in the Wolverhampton direction towards the tunnels under Livery Street and Northwood Street. The coaching stock behind 2903 is in New Yard sidings. 2903 was already twenty-three years old at this point and would survive until November 1949 being withdrawn from Tyseley. *(George Barlow)*

The Great Western Railway's contribution to streamlining in the mid 1930's sees 6014 *King Henry VII* at the south end of Snow Hill station on an up working. Modifications were the bulbous smokebox door, fairings behind the chimney and safety valve, and the raked front to the cab along with the single wheel splasher which was seen again on Hawksworth's Counties post war. The fairings in front of the cylinders have already been removed. 6014 was allocated to Wolverhampton Stafford Road in 1934 and it would be withdrawn from there in September 1962. Above the engine is the footbridge with walkway from the booking office above the [illegible] (*J A Whaley*)

The Great Western Railway were early proponents of diesel traction with the AEC diesel railcars. Introduced in 1933 No. 8 entered traffic in 1936. These units were used on a Birmingham Snow Hill to Cardiff service which was the first long distance diesel express service in Great Britain. With only 44 seats, capacity was limited. Parcels vehicles can be seen in the bay platforms 9 and 10. *(Flint & Harbart)*

'Aberdare' 2656 sits on the Down through road at Snow Hill on a mixed goods. Allocated to Pontypool Road in 1934 this may be on a working bound for the goods yard at Hockley a few miles towards Wolverhampton. A curious William Dean design with outside frames and a very recognisable Swindon boiler 2656 would survive until March 1948 after a working life of over forty-six years. *(George Barlow)*

A lovely scene inside the overall roof at Birmingham Snow Hill 3278 was built in 1897 and would only survive until December 1938. The loco along with Bulldog 3436 formed Dukedog 9022. The original Great Western Station at Birmingham Snow Hill opened in 1852. After rebuilding, the classic Snow Hill seen here was opened in 1912. Constrained by Livery Street on the west side and Snow Hill itself on the east, the station featured two 1200 foot long platforms with bay platforms at the North end. The decorative features of the platform buildings at Snow Hill can be seen with salt glazed wall bricks and buff terracotta cornices strings and copings can be seen along with the large station clock showing 3:30pm and the Refreshment room and Luncheon room. *(Transport Treasury)*

LNWR Prince of Wales Class 25601 *Andromeda* outside Aston Shed 15th July 1934. Built at Crewe Works as w/n 5031 and entering service in October 1911 she was withdrawn in March 1935. Designed by Charles Bowen Cooke, 245 of the class were built between 1911 and 1919, being the LNWR equivalent of the later Stanier Black Five, *Andromeda* being the second of the class built. Six would survive into BR service. Aston was the principal LNWR shed in the Birmingham district and opened in 1883. Coded 10 by the LNWR and then 3D by the LMS in 1935, Aston was a twelve road shed and closed in October 1965. *(Neville Stead Collection)*

A wonderful view of Navigation Street with Birmingham Corporation Tramways car 522. To the left is Queens Hotel and then behind the overall roof on the LNWR side of New Street Station. Route 37 was the Cannon Hill service to Willows Road Edgbaston and would survive as a Tram route until October 1949. Car 522 would survive a little longer until October 1952. Hidden behind the tram was Finlay's tobacconist and beneath the Swan Vestas advertisement is the awning for waiting tram passengers. The roof, which on construction was the largest in the world with a span of 212 feet was badly damaged in an air raid on the 26th October 1940. The scaffolding on the apex of the roof used to hold advertising, the 'News of the World' featuring in pre war images. The board to the right is for Horatio Lane, Byrne & Ralphs auctioneers and were based in Colmore Row. Trams ceased running in 1953 but returned to Birmingham city centre in 2015. *(Frank Goudie)*

Webb 2P 'Watford' tank 6927 was built at Crewe Works in 1902 carrying LNWR number 2211. Renumbered by the LMS she became 6927 in December 1927 and survived into BR service being withdrawn in May 1948. It is on shunting and pilot duties on the LNWR side of Birmingham New Street in what I think are the former Stour Valley bays. First vehicle is an open carriage truck with a pantechnicon trailer marked with 'on tow' upon it. The overall roof that once covered this former LNWR side of Birmingham New Street started to be dismantled in 1945 taking 2 ½ years to complete, some of the remaining roof spans visible top right, the station being rebuilt with functional canopies which we will see later. Queens Hotel dominates the backdrop. *(Frank Goudie)*

Midland Railway 483 Class no 517 is seen on the Midland Railway side of Birmingham New Street. A notional rebuild of an earlier Johnson 4-4-0, 517 would be withdrawn in June 1949, never receiving her BR number. Allocated at Bournville shed at Nationalisation. Sat in the Fish yard sidings with Hill Street in the background. A one horsepower shunter and dray sit alongside. The sidings were built as part of the 1885 expansion of the station to serve the nearby Birmingham Fish Market, with it written that vans would arrive from Hull, Aberdeen and Fleetwood. This area of the station was noted as having a pervading smell of fish. The metalwork carries a glazed canopy over the area. *(J T Clewley)*

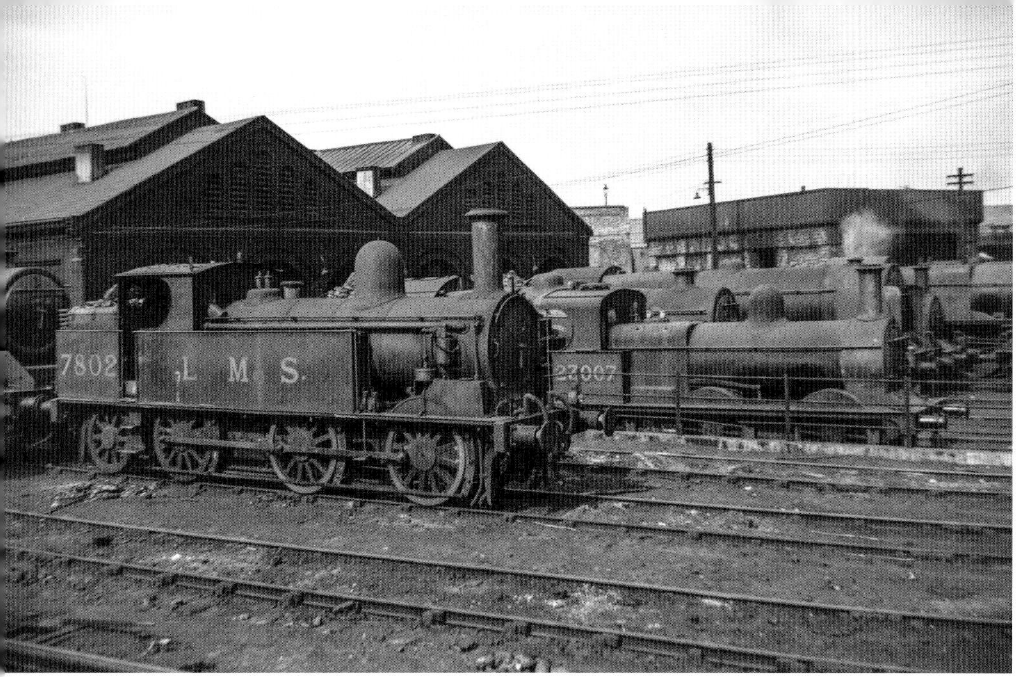

Left Top: A fascinating scene at Monument Lane shed just post nationalisation. LNWR Coal tank 7802 sits in front of a Stanier Black Five and was a new arrival coming from Plodder Lane in November 1948. Its stay would be short being withdrawn in October 1949. Johnson 2F 23007 sits behind. Built by Neilson Reid & Co in 1876 the engine would soon be renumber 58179 and would continue in service until April 1956 after almost eighty years in service being scrapped at Derby Works. Tantalisingly there are certainly one, possibly two parallel boiler Royal Scots in the scene. Situated on the LNWR Stour Valley line between Birmingham and Wolverhampton, Monument Lane engine shed opened in 1858. In the 1930's the LMS added mechanical coal and ash plants alongside a 60ft turntable and three additional stabling roads. Coded 11 by the LNWR the shed became 3E in 1923, 21E in 1960 and finally 2H in 1963. Final closure came in December 1966.
(Transport Treasury)

Left Bottom: Rebuilt Patriot 45514 *Holyhead* is emerging from the tunnels and under Hill Street and Navigation Street bridges and into Birmingham New Street from the Wolverhampton direction on a London bound express. It was Bushbury 3B allocated before moving to Carlisle Upperby in May 1949. The tender still carries LMS, but the engine now has its post nationalisation number. Built at Crewe Works in 1932, the engine was rebuilt in March 1947.
(Flint & Harbart)

Right: LNWR 'Prince of Wales' 25752 in Birmingham New Street. Rebuilding of the station post removal of the overall roof is ongoing. The canopies were intended to be temporary pending a new overall roof but would remain until rebuilding of the station in 1960's. At nationalisation 25752 which was an unnamed member of the class was Stafford allocated so this may be on a Stafford local service.
(Flint & Harbart)

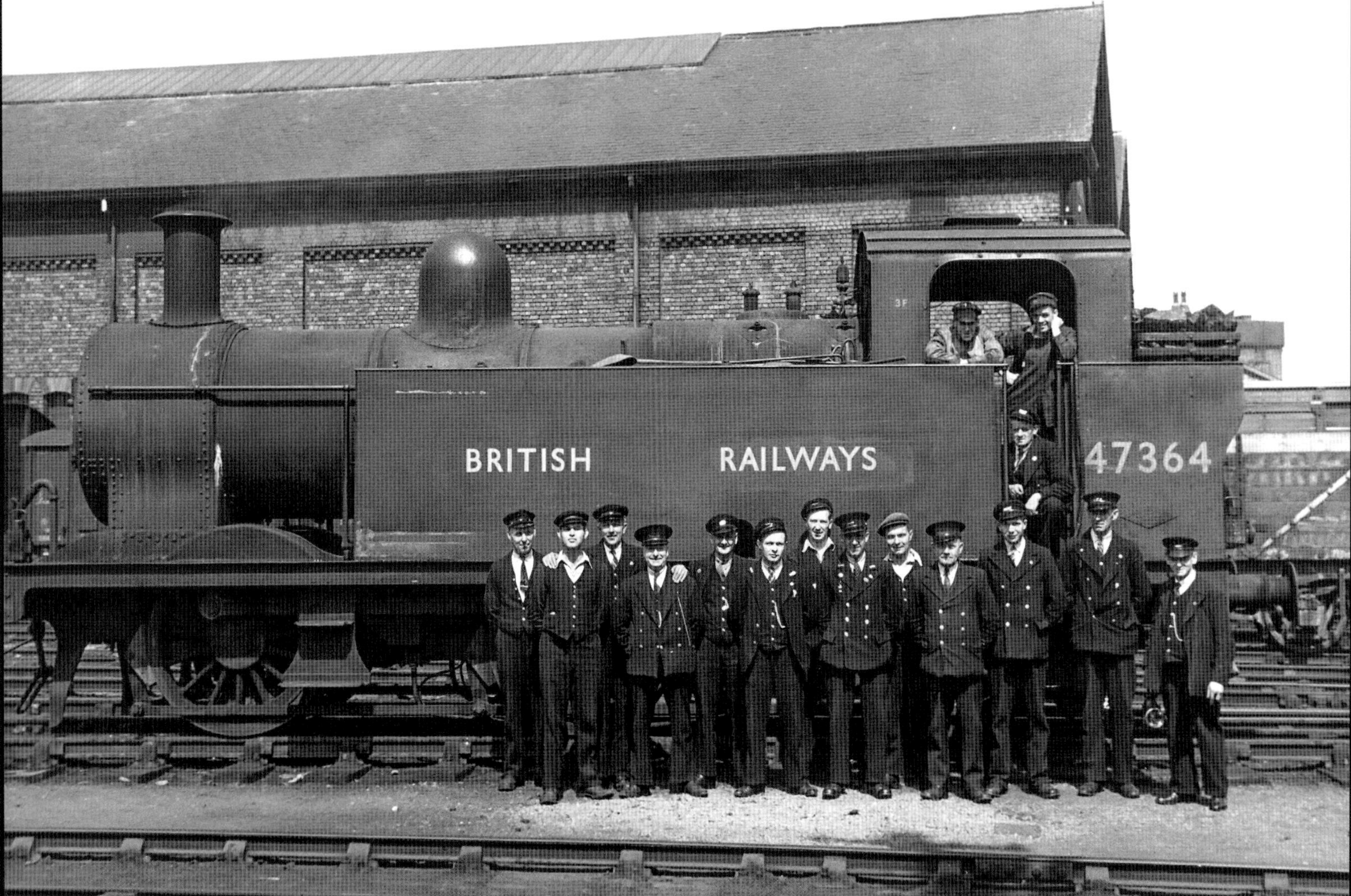

Just post nationalisation is Fowler 3F 47364 at Curzon Street. Allocated to Aston 3D she carries the early Gill Sans British Railways lettering before the more recognisable lion and wheel emblem. Following its closure as a passenger station Curzon Street became a focal point for goods traffic in and out of Birmingham. By 1914 nearly 2000 men were employed at Curzon Street and in 1923 over 1,500,000 consignments handled. There were 8 warehouses with almost 130,000 square feet of floor space and 820 wagons could be positioned for loading and unloading. It was a cramped site with many short sidings which made working complex and time consuming. *(J C Flemons)*

Midland Railway 700 class and Kirtley designed 22846 on 21B Bournville shed. Built at Derby in 1873 this locomotive would be finally withdrawn in September 1949 after seventy-six year's service and be the last of its type to remain at Bournville. Behind sits fellow Bournville allocated Stanier 3P tank 173, a much younger engine entering traffic in January 1938. Bournville shed was a standard Midland Railway roundhouse opened in 1895 and was situated south of Bournville station and alongside the Lifford Canal branch junction. A little further south is Lifford West Junction which gives access to the Camp Hill line. Bournville was home to a number of these outside framed engines which were mainly used on the Halesowen branch. In the background can be seen the coaling stage. Closure of Bournville shed came in January 1960. *(Milepost)*

A short line ran from opposite Bournville shed down to sidings alongside the Worcester and Birmingham canal serving a number of canal wharves. Transhipment of goods had long ceased in this 1959 image. Originally part of the Birmingham West Suburban Railway line, and conceived to support development to the south of Birmingham the line ran from a terminus at Granville Street (close to Five ways) and followed the canal passing under the Birmingham and Gloucester railway viaduct seen here before climbing up and joining this line at Kings Norton. The potential of the BWSR as a through route through Birmingham New Street saw a more direct route from Bournville to Kings Norton constructed. With the opening of the Midland Railway line through New Street the original Birmingham and Gloucester Railway which ran from Kings Norton, became a useful route for through goods traffic on the Midland avoiding Birmingham New Street along with circular suburban services through New Street, Bournville, Kings Heath and Moseley. The suburban passenger services were withdrawn in 1941 and the stations closed; however at the time of writing new stations are being constructed with the aim of reintroducing passenger services along the route. *(A E Bennett)*

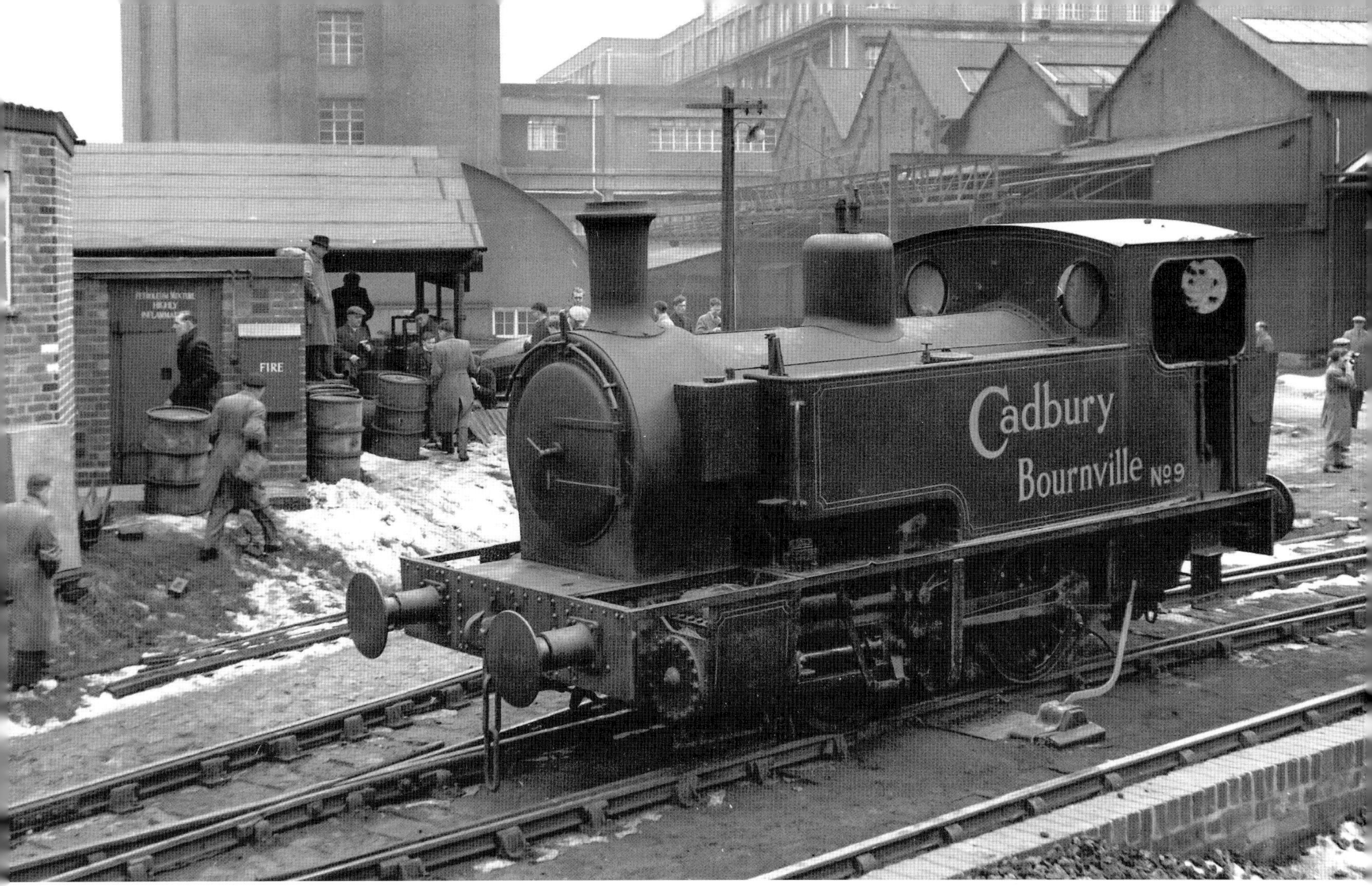

We cannot leave Bournville without a look at the Cadbury Chocolate factory internal rail system. The Cadbury family established their chocolate factory just to the west of what at the time was known as Stirchley Street station (but would become known as Bournville) in 1879. Initially deliveries from the canal and railway were made by horse and cart but growth soon made this unsatisfactory and so in 1884 the works railway opened. At its height it totalled six miles circling the whole manufacturing area and crossed over both the Birmingham West Suburban Railway and the canal alongside to give rail access to the canal side warehouses. On what looks like a society visit Hunslet w/n 3665/1949 No.9 is seen with the factory as a backdrop. The last trains ran on the internal network in March 1976 and Cadbury No.1 is in preservation at Tyseley. *(Flint & Harbart)*

Stourbridge Junction 84F allocated Collett 74xx sits at Longbridge station. Situated on the former Halesowen Railway which ran from Old Hill to a junction on the Midland Railway Birmingham to Bristol main line the line became joint GWR and LMS operation. The station served the Austin motor car works seen behind and still sporting its wartime camouflage paint. The main station building was on the road bridge (which is out of shot to the right). Longbridge didn't appear in the public timetables and was served by workmen's trains running from either Old Hill by the GWR or Birmingham New Street and run by the Midland Railway and later LMS. Services were withdrawn in 1958 and 1960 respectively. 7449 was new to traffic in April 1950 and would be withdrawn in June 1963. *(Neville Stead Collection)*

The Austin motor company had an extensive internal rail network and its own fleet of locomotives including two USATC S100 class built by Davenport locomotive works in 1943. This example is USATC 1938 w/n 2505 named *ADA II* acquired in 1947 and reported scrapped in 1957. Other former Austin engines have fared better with Bagnall built *Vulcan* and *Victor*, Kitson built *Austin No. 1* and Manning Wardle *Abernant* all surviving into preservation. The Longbridge works built cars until 2005 closing with the collapse of the Rover Group, with the site now cleared and redeveloped. *(Neville Stead Collection)*

An SLS special from Birmingham New Street to Harborne is seen here passing Monument Lane. The Stour Valley line to Wolverhampton emerged from the tunnels at Monument Lane passing Monument Lane engine shed and then going under St Vincent Street which can be just seen in the background. The tour which used a two coach Dudley Port push pull set was the first passenger train to Harborne since 1934 and was run on the agreement no passengers would alight. The two and a half mile branch was opened in 1874. Competition from shorter bus routes into the city saw passenger services withdrawn on the 26th November 1934. Freight services would continue until 4th November 1963 when the branch closed. LNWR 1P 46757 entered traffic in July 1897 from Crewe Works and entered traffic as LNWR 1478. Allocated to 3C Walsall Ryecroft, reallocation away from the Midlands to Sutton Oak followed in November 1950. She would survive to February 1953. Having visited Harborne the tour would take a circular route back to New Street via Perry Barr and Stechford. The fare was 4/6d and the tour left New Street at 15:10 returning at 16:55. *(Milepost)*

The SLS returned to Harborne on the 30th May 1959 with a rail tour which ran topped and tailed with Midland Railway Johnson 2F 58283 and 58271 with a set of six Stanier coaches. The station at Harborne is shown along with the goods yard and unlike the 1950 tour passengers were allowed to alight, albeit not on the platform! The building above the coaches is the Chad Valley Factory, well known for tinplate toys. To the left can seen the goods shed and yard. *(Henry Priestley)*

Below: Southbound workings from Snow Hill and Moor Street crossed the long sixty-three arch viaduct to Bordesley where we see an unidentified Churchward Mogul on a long goods working. To the right can be seen the cattle pens, Birmingham livestock market was in nearby Digbeth, so the pens were kept busy with traffic for the regular livestock markets. Unlike New Street where freight could take avoiding lines via Camp Hill or Perry Barr and Stechford, all freight heading through Birmingham from the Wolverhampton and Tyseley directions had to pass through Snow Hill. *(J C Flemons)*

Right: Duddeston Viaduct diverged at Bordesley and was built to carry the GWR promoted, Birmingham and Oxford Junction Railway to a connection with the Grand Junction at Curzon Street, This would allow the GWR access to a direct route to Liverpool and the Grand Junction Railway an alternative route to London than the London and Birmingham Railway. The amalgamation of the London and Birmingham and Grand Junction to form the London and North Western Railway changed matters somewhat and the GWR submitted an extra bill for the Birmingham Extension Railway from Bordesley to Snow Hill. However, the original bill was for access to Curzon Street and the viaduct had to be constructed and still stands. It was used as a head shunt to the yard at Bordesley and stabling cattle wagons for cleaning. *(J C Flemons)*

Tyseley Shed and sat under the wartime ash shelter is Oxford 81F allocated Dukedog 9028. In the background the roundhouse can be made out, one of two at Tyseley. 9028 entered traffic in November 1939, the last of the class built. Lurking behind is a Stanier Jubilee recorded as 45721 *Impregnable* which was 8A Liverpool Edge Hill allocated and was recorded at the time as having a Fowler tender. Given the reporting number just visible on the smokebox it had maybe worked an excursion to Stratford-upon-Avon. *(Transport Treasury)*

One of the British Railways built batch of Collett Manor's 7821 *Ditcheat Manor*. Built at Swindon in November 1950 she was Tyseley allocated from December 1953 to July 1959. Tyseley roundhouse is just out of shot to the left and visible behind 7821 are a number of early diesel shunters. 7821 is carrying the early lined black livery but would soon be outshopped in lined green. In Rail 200 year the Manor presence at Tyseley is still retained with 7822 *Foxcote Manor* under overhaul at Tyseley Locomotive Works and 7802 *Bradley Manor* and 7812 *Erlestoke Manor* have been outshopped. *(Transport Treasury)*

Leamington Spa allocated Large Prairie 5185 is seen approaching Acocks Green and South Yardley Station on a Birmingham bound local. The leading four coaches are a Birmingham division suburban set, made up of two brake third and two composites. In the Birmingham division these were termed B Sets, the more familiar two coach sets termed B sets elsewhere becoming D Sets. Early batches were Toplight stock with later diagrams being bow ended 58 foot long, nine foot wide coaches and a final batch 61ft long. These were close coupled via a draw bar with no buffers except on the outer ends. Travelling on the down main line, with the London bound up main to the left and then up relief which extended as far as Lapworth and down relief on the far right. *(J C Flemons)*

Moving to the LMS lines out of New Street we look east at Proof House Junction. To the right and under the flyover the lines to Coventry and Derby. The flyover carries the line to Aston with lines on the left giving exits from Curzon Street goods yard to both Aston and Coventry/ Derby directions. Curzon Street No.1 Signal box, a LNWR pattern example controls the lines out of the goods yard. Gloucester Junction would follow next with the Midland Railway Line to Derby diverging along with a connection to the line to Gloucester via Camp Hill. *(J C Flemons)*

Saltley 21A and resident Fowler 4F 44084 sits on shed. Built by Kerr Stuart she entered traffic in October 1925. She received her BR number in Feb 1949 so being reasonably clean probably dates the image to not too long after. Saltley was the largest Midland Railway depot after Derby and featured three roundhouses. Coded 3 by the Midland Railway, the shed was recoded to 21A in 1923 and then 2E in September 1963. All three roundhouses were re-roofed post war and No.3 roundhouse can be seen being rebuilt in the background. In 1954 there were 191 engines allocated to Saltley. *(Flint & Harbart)*

Ancient and modern at Saltley. Horwich built Ivatt 4MT 43014 which was new to traffic in April 1948 and moved to Saltley from Sheffield Grimesthorpe 19D in December 1948. The sojourn in the Midlands only lasted until when she moved north to Stourton. Carrying the early British Railways livery 43014 also carries a double chimney the early members of the class were built with. Nicknamed doodlebugs in double chimney form they had a reputation for indifferent steaming, and all would become single chimney. In the background sits Kirtley 700 class 22853. Built at Derby in 1873 she would be withdrawn from Saltley in March 1950. In the background being re-roofed is Roundhouse no.3. *(Flint & Harbart)*

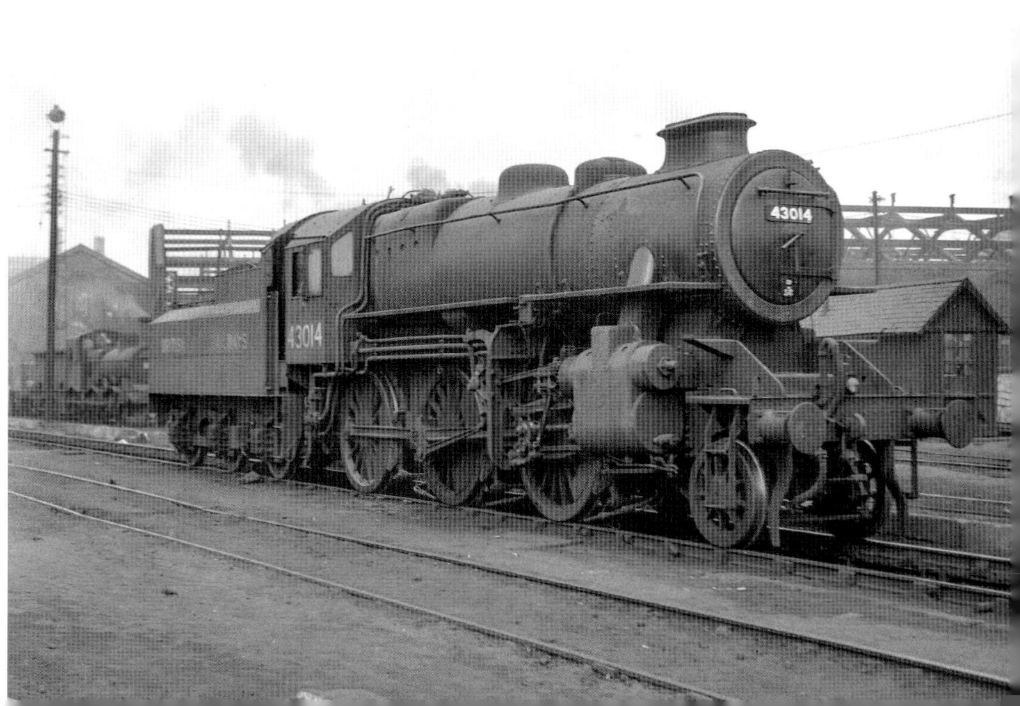

Post war progress with LMS Diesel shunters are seen at Saltley on June 13th 1948. Introduced in 1946 over 120 would go on and be built with 7126 built at Derby entering traffic in August 1947, her first home being Saltley where she would stay until October 1966. Designated class 11 by British Railways 7126 would become 12039 in August 1952 and would remain in service until October 1968, being withdrawn from 5A Crewe Diesel depot. Behind is Fairburn 4P 2-6-4T 2685 which had a number of periods allocated to Saltley. The background is office and locker rooms with the water tank above, along with stores and offices and Engine Shed no. 2. *(J S Cockshott)*

Moving to the west end of New Street Station we see Compound 41089 departing from the Midland side of Birmingham New Street. Built at Horwich and entering service in June 1925, her final allocation would be to Bournville 21B from Leicester 15C in Jan 1957 with withdrawal coming in July 1957. 43668, a Johnson 3F which entered traffic in 1900 was a Bournville engine at Nationalisation moving to Saltley in February 1960 and finally being withdrawn in August 1962. Navigation Street forms the backdrop and behind the 3F was a turntable. The foreground footsteps were for Birmingham New Street No. 5 signal box. *(Harold Cartwright)*

NTLEMEN

Left: The growth in traffic by the early 1870s on both the LNWR and Midland Railway led to the expansion of New Street. The LNWR acquired land south of Queens Drive with construction starting in 1881. The work was done by the LNWR with the Midland effectively a tenant and added an additional four through platforms. The platforms were covered with a semi-circular curved roof glimpsed here. Great Queen Street which ran through the middle of the station and separated the Midland and North Western sides can be seen in the background. A father and son observe the cameraman from the lead droplight in a Stanier porthole carriage. *(Transport Treasury)*

Below: Selly Oak was opened in 1876 on the Birmingham West Suburban Railway. Initially single line the doubling took place in 1885 with the extension from the west into New Street. Saltley 21A's Stanier Black 5 45006 passes through on a southbound stopping passenger four coach set of Stanier vehicles. The top lamp bracket has been moved to the side of the smokebox with electrification ongoing. The station was in the middle of sidings with a goods yard to the right tantalisingly out of shot. The main station building is on the left and the steps down to the underpass under the line and out onto the Bristol Road can just about be glimpsed. *(Arthur Mace)*

Near what is now University station Stanier Black 5 44666 heads a southbound working away from Birmingham with a rake of Stanier coaches. 44666 was a long term Saltley loco arriving in November 1949 and leaving in October 1965 when it moved to Tyseley. Alongside is the Worcester and Birmingham canal which the Birmingham West Suburban railway runs close to from Fiveways and out to the south side of Bournville. *(Arthur Mace)*

The Stephenson Locomotive Society Birmingham Area Rail tour which we saw at Harborne behind steam earlier swapped steam for a Metropolitan Cammell Class 101 DMU for the second part of the day's tour and seen here at Suffolk Street Goods station which was close to the original terminus of the Birmingham West Suburban Railway at Granville Street. After the Second World War passenger rated parcels traffic was moved from New Street to here. A mixture of British Railways and earlier LMS parcels vehicles can be seen on the left. *(Transport Treasury)*

The Midland side of Birmingham New Street with Saltley's Stanier Black 5 45265 seen alongside Birmingham New Street No. 2 signal box. The curved roof of the Midland side in comparison to the long span of the LNWR train shed is apparent. In the background can be seen stock in Platform nine. *(Arthur Cundall)*

A first generation Derby Lightweight two car DMU sits in Platform four. The very austere canopies from the post war rebuilding are apparent. Queens Hotel overlooks the scene. The speed whiskers of the early DMU liveries before half and full yellow ends was quite striking. Introduced in March 1956 these units were used on the Birmingham to Lichfield and shorter Four Oaks services but would be replaced by May 1958 with Metro Cammell units, helping to date this scene. These units were non-standard and all withdrawn by 1969. *(Harold Cartwright)*

Introduced in 1947 and built at Derby Works 10000 was the first of two main line diesel locomotives introduced by the LMS. These 1600hp machines were designed to operate express services as twins but semi fasts and freights singly. Originally outshopped in a striking black and silver livery 10000 is seen at Birmingham New Street, now in British Railways green and carries the late crest and would survive until 1963. Sister engine 10001 would be withdrawn in 1966. *(Transport Treasury)*

The atmosphere of a steam era roundhouse, in this case Saltley No.3. Of the engines on view 92205 was a 50A York engine and the three Stanier engines Saltley allocated. Saltley became 2E on the 1st September 1963. The glass frontage of the post war rebuilding is apparent behind the engines.
(Chris Lodington)

The background is Saltley Engine shed No. 3 post its rebuilding. To the right is the chimney for the sand house and lined up in front three Riddles 9F 92165, 92166 and 92167 all built at Crewe in 1958 and fitted with mechanical stokers to ease some of the workload on the fireman but also to use lower grade coal. Not entirely successful the engines were all converted to normal firing. To the left a rake of 16t mineral wagons sits, some with doors open. Note the discarded ovoids in the foreground, fire shovel, brazier and trolley all just lying on the ground. *(Alec George)*

41D Canklow allocated Stanier 8F 48765 is seen at Saltley station. Doncaster built and entering traffic in March 1946 as LNER 3160 she became LMS 8765 in October 1947. Her stint at Canklow was only from May 1961 to December 1962. Looking towards Derby the cooling towers of Nechells Power Station can be seen as can the gas holders at Washwood Heath. Nechells gas works are out of shot to the left. Saltley sidings signal box can be seen to the left which closed in September 1969. Saltley station was an island platform opened in 1899 albeit replacing an earlier station and closed in March 1968 with the withdrawal of the Derby-Tamworth-Birmingham local services. To the right are the Saltley works of Metropolitan Cammell which closed in 1962, and activity concentrated at Washwood Heath. *(David Moore)*

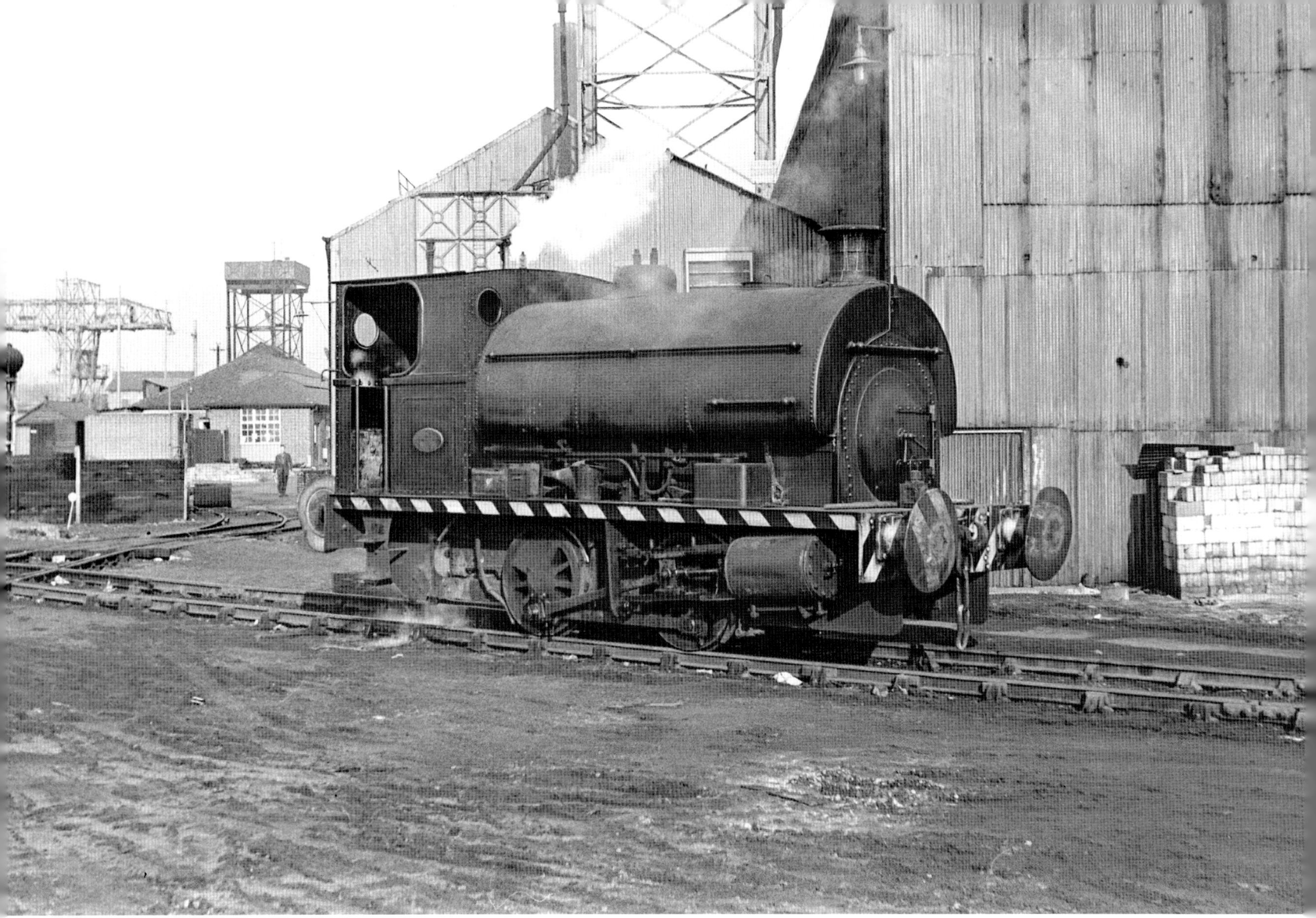

The Midland line to Derby on its exit from Birmingham ran through a very industrial area which developed along the River Rea. Dunlop works, Bromford Tube works (seen here), Gasworks and the power station at Nechells each had their own internal user networks. Extensive sidings were developed by Washwood Heath with many trip workings, across to the GWR at Bordesley. Bromford tube works opened in 1922 alongside the Birmingham to Derby Line with the exchange sidings alongside Bromford Bridge station. Peckett W6 class w/n 1936/1937 is seen here inside the works. Withdrawn in 1967 she would be scrapped on site. The works themselves closed in April 1994. *(Alec Swain)*

Whilst we have grown used to our electricity and gas supplied nationally at the beginning of the 20th century provision was more local with the City of Birmingham Electricity supply department having the responsibility for electricity provision in the city and surrounding area. To cope with rising demand for electricity, Nechells was identified as a location and construction of the 'A' power station started in 1919 with electricity generation from June 1923. A second 'B' power station was approved in 1945 and opened in stages from 1951 to 1953. Robert Stephenson and Hawthorn 7684/1951, Nechells No.4, is seen here at Nechells in this undated scene. In the background a rake of 20t coal wagons can be seen. *(Milepost)*

The station at Castle Bromwich opened in February 1842 as part of the Birmingham and Derby Junction Railway direct line into Birmingham. When the line was quadrupled the goods lines ran at the back of the two platforms. The main station building was to the right and behind the photographer standing on the south side of all the running lines and fronting onto Chester Road. The British Industries fair took place at Castle Bromwich from 1920 through to 1960 generating a significant amount of traffic both in excursions but also shuttles to and from Birmingham. A three car Derby Lightweight DMU is seen on a Birmingham bound service. Just beyond Castle Bromwich station were the connections to the line through Sutton Park to Walsall. *(Transport Treasury)*

We now visit Bescot which opened in 1892 to serve the sorting sidings, also opened in that year. By 1954 the allocation was seventy engines, twenty seven of which were Stanier 8F's. The shed had eight roads and under the LMS was 3A and concentration depot for the western division Birmingham area. By 1963 Bescot had become 2B and would end the steam era as 2F. Steam duties ceased in March 1966, but Bescot would remain a diesel maintenance facility. Two Stanier 8F's are seen as part of a long line of engines including a Super D, Ivatt Class 4MTs and an Ivatt 2MT. *(Alec George)*

It's the tail end of steam at Bescot and Fowler 4F 44139 is seen with Bescot station in the background. Only allocated to Bescot in February 1965, she would survive until August. The original LNWR station buildings have been demolished and the station is in the throes of rebuilding, along with electrification masts being erected. The marshalling yard was also remodelled at this point as well. The backdrop would soon change with the construction of the M6 motorway. The cooling towers are from Birchills Power station at Walsall as well as the Gas holder at Walsall gas works. *(Trevor Davis)*

Wednesbury on the 28th May 1960 and LNWR Super D 49275, which was Bescot allocated from December 1958 until withdrawal in October 1961, is seen shunting a number of tarpaulined open wagons. The lines in the foreground are the South Staffordshire Railway from Dudley to Lichfield and Burton on Trent via Walsall, this section opening in 1850. The embankment the photographer is stood on carries the GWR Birmingham Snow Hill to Wolverhampton main line. The curved sidings extended round to a connection with the GWR line and just out of shot to the right were Brunswick sidings alongside the GWR line. In the background is Ocker Hill Power station which closed in the early 1980's. *(R C Riley)*

The view of the South Staffordshire Railway, Wednesbury station captured from the GWR line as it crossed over. Opened in 1850 the station would become Wednesbury Town in 1950 and close in 1964. The GWR station survived a little longer closing in 1972. Wednesbury No. 1 was an LNWR pattern signal box and also controlled access to the sidings we saw with 49275 earlier. In the background is Darlaston Junction and Darlaston loop which doubled back to a junction on the Walsall to Wolverhampton line. St John's Church dominates the skyline whilst the houses on Stafford Street look on over the railway. The level crossing by Wednesbury No.1 carries Potters lane over the railway. *(R C Riley)*

A look now at Wednesbury Central station on the GWR line to Wolverhampton, similar in design to Handsworth. The connection to Brunswick Sidings and the South Staffordshire Line is out of the image on the left. Hawksworth County 1004 *County of Somerset* is seen arriving on a what is suggested to be the 8:55 Birkenhead to Paddington Parcels working. Leading vehicle is a Fruit D, then a British Railways Mark one full brake. In the background is an Iron and Steel foundry, once so much part of the industrial landscape of Birmingham. Just above the cab can be seen the GWR type 7 signal box which was situated on the platform. *(R C Riley)*

The GWR Kings had enjoyed something of a main line swansong on the Wolverhampton to Paddington expresses and with the class coming to the end of their service life, on the 17 November 1962 LCGB ran a rail tour to mark the occasion with a non-stop run from Paddington to Birmingham before progressing to Wolverhampton and is seen here at Hockley. Passengers enjoyed a much more circuitous return to Marylebone with firstly a GWR 66XX and then a 2251 class. 6018 *King Henry VI* would be withdrawn in December 1962 but reinstated in April 1963 for the Farewell to the Kings tour.
In the background is the GWR Hockley goods depot, the principle GWR depot in the area and opened in 1854. The station was unusual in having no platform for the down main line, the island platform seen here serving the local lines and the up platform for the main just visible on the left.
The station would close in 1972. *(Alec Swain)*

The Birmingham, Wolverhampton and Dudley Railway opened in 1854 and Handsworth and Smethwick was one of the intermediate stations. A Wellington bound DMU is seen departing from the down local platform. The station buildings, and those at Bilston, Wednesbury and West Bromwich were of the same similar classical but slightly austere style. The station would close in 1972. *(R C Riley)*

The main platform roof at Snow Hill ran from Great Charles Street, which goes underneath the station to the footsteps this image was taken from. Collett Castle 5063 *Earl Baldwin* arrives at Platform 7 on an up working at Snow Hill in this early 1960's scene. With the clock on 12:30pm this may be the late running 8:50am from Birkenhead due out of Snow Hill at 12:00 and into Paddington at 2:22pm. *Earl Baldwin* was Wolverhampton Stafford Road allocated from April 1958 to September 1963. So much is to be noted in the image from the Wymans bookstall to the circular station totems. The Pullman car B sign is for the Birmingham Pullman service introduced in September 1960. *(Arthur Mace)*

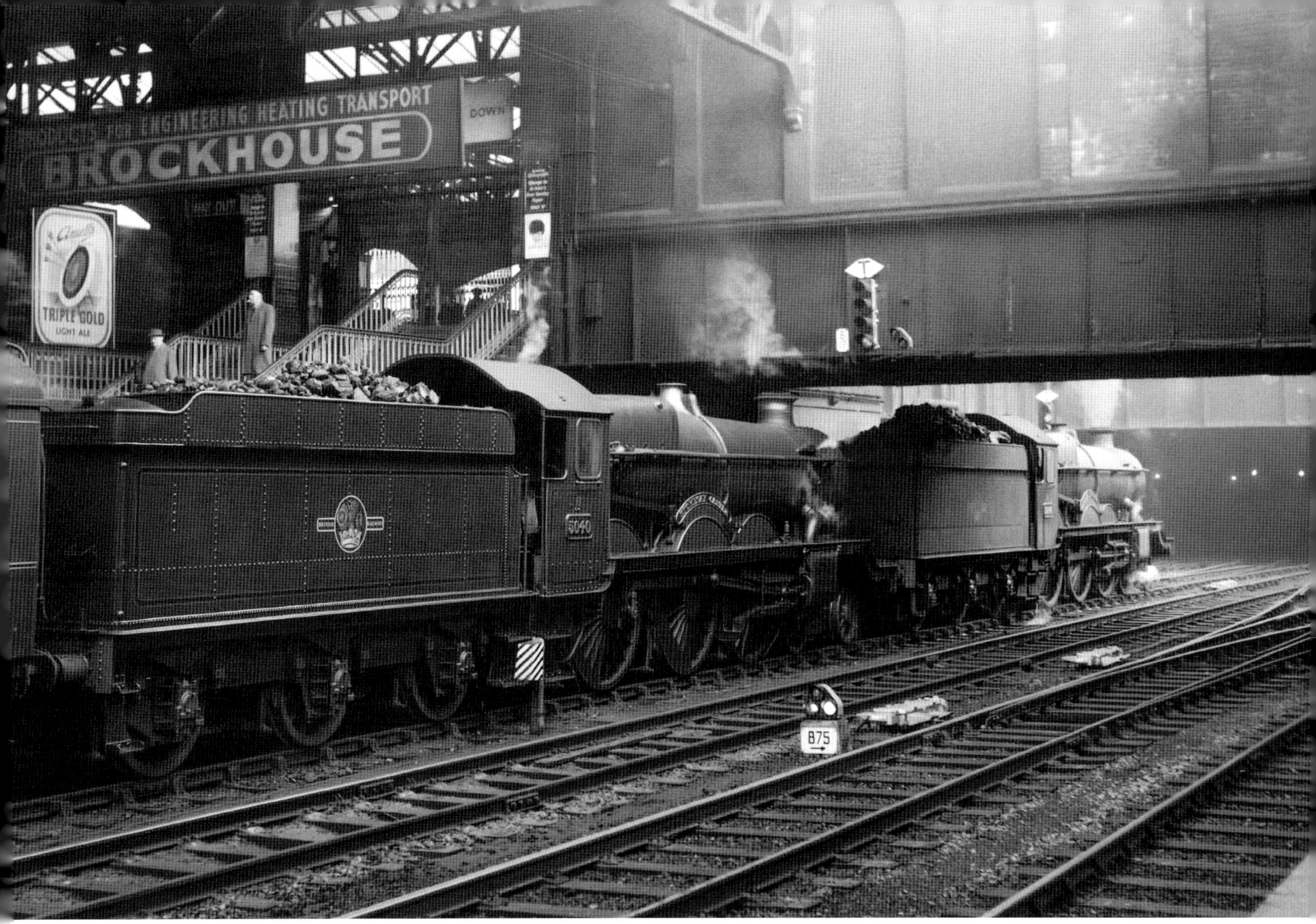

The glory of Snow Hill in the latter days of steam and two Castles at the head of an up working. 5040 *Stokesay Castle* was on a brief allocation to Swindon before moving to St Philip's Marsh at Bristol and then being placed into store in February 1962 and 5049 *Earl of Plymouth* already a St Philip's Marsh engine. Behind the loco are the steps leading up the footbridge and station concourse. *(Arthur Mace)*

A magnificent view of the north end of Snow Hill taken from the signal box. Livery Street to the right and Snow Hill to the left. The long platforms for the through lines were split into two, so platforms 1 and 2 alongside Livery Street and the bay platforms 3 and 4 are occupied by DMUs. Stafford Road allocated 6001 *King Edward VII* is seen departing from platforms 5 and 6 with a Paddington to Birkenhead train. Almost hidden by the first coach a GWR large Prairie sits on a rake of suburban coaches. To the left further non corridor coaches of both GWR and BR design sit on the up-loop siding. Above the overall roof can be seen the roof line of the Great Western Hotel which fronted onto Colmore Row. With the closure of the GWR main line through Snow Hill in 1967 along with closure of the station itself in 1972 the rails were removed and station demolished. The third iteration of Snow Hill reopened in 1987 with a car park replacing the overall roof. New office blocks on Snow Hill tower over the scene today. *(R C Riley)*

The decline of steam on the Western Region in the Midlands. Shorn of name and numberplates 6848 *Toddington Grange* approaches Acocks Green and South Yardley on the Down Main line. 85A Worcester allocated the engine would survive until the end of steam on the Western Region on 31st December 1965. The signal box and slow lines are no more, replaced with a car park. *(Arthur Mace)*

GREAT WESTERN

Left: Birmingham became the birthplace of the modern preservation movement. GWR small prairie 4555 was withdrawn in November 1963 and purchased for preservation by Pat Whitehouse and Pat Garland for the Dart Valley Railway but her initial preservation home and close to the homes of the purchasers was Tyseley. In return for her accommodation, she was used on light local passenger and freight duties with the 5:27pm semi fast to Dorridge a regular turn. Seen here on the coaling stage road at Tyseley shed. One of the two roundhouses in the background and an unidentified 8F and Hall can be seen. Towards the end of steam, Tyseley had a number of Stanier 8Fs allocated. The carriage sidings can be seen in the background and beyond that is Tyseley station itself. The descending lines are the goods loops. Just above the cab of 4555 can be seen a number of sidings which as we shall see shortly served a scrapyard. Maps of 1951 suggest the imposing building was a hostel. *(R C Riley)*

Above: From January 1963 Snow Hill became part of the London Midland Region and eight years later and a complete contrast to the earlier image Snow Hill as a station is no more. March 1967 saw express workings cease and March 1968 Snow Hill Tunnel closed with local services from Dorridge and Stratford terminating at Moor Street. Snow Hill became a car park and a selection of classic vehicles are on display on the platform and former running lines. *(Bernard Mills)*

One service lingered from Snow Hill with a poorly advertised service to Wolverhampton running from the bay platform. Livery Street is to the right with the glass screen missing glazing and the end of the trainshed visibly in disrepair. This service would cease in March 1972. One wonders where the Way Out to Great Charles Street enamel sign is now. *(Bernard Mills)*

Our first look at Birmingham Moor Street, built by the GWR to increase suburban capacity on the lines to Lapworth and Stratford upon Avon. Originally opened in 1909 the station you see here opened in January 1914. Note no points for a run round, this being enabled by a traverser which ran under the platforms. The Rotunda towers over the old goods depot which itself had wagon lifts to lower wagons underneath for unloading. The depot itself increased capacity and with close location to the produce markets of central Birmingham reduced the need for workings through Snow Hill for unloading at Hockley and then transhipment across the city. The depot would close in 1972 and the shed demolished to create a car park. *(Bernard Mills)*

Below: With the end of steam on British Railways in August 1968 and an embryonic preservation movement, the open days held at Tyseley attracted tens of thousands of visitors, this one on 3 October 1971, over 18000. Steam had just returned to the main line with the test workings by 6000 *King George V*. Stanier Pacific 6201 *Princess Elizabeth* is seen at one end of shuttles which ran along the goods loops at Tyseley. Wharfedale Road bridge is packed, as is the platform and the lineside on the right. *(G H Taylor)*

Right: New Street was once again rebuilt in the early 1960's and this is the Midland side, the old train sheds have been removed along with Queens Drive. A concrete deck covered the platforms with the concourse above and in time the Pavillions Shopping centre. The station was still partially in use with the set of coaching stock in one platform. St Martin's church is seen alongside the Bull Ring and its shopping centre and out of sight the open air markets. St Martins still stands but the Bull Ring seen here redeveloped as is the space above New Street, which is now Grand Central. *(Arthur Mace)*

The resulting New Street was a very concrete, dark and at times smoky station but it had a charm. This is the London end of the station on 7 February 1975. The Paddington trains which used to go to Snow Hill were diverted to New Street, bringing Western diesel hydraulic D1064 *Western Regent* at the helm. It was a Plymouth Laira engine at the time and would be withdrawn from there in December 1975. Alongside an AM10, later class 310 four car electric multiple unit which from the destination blind is bound for Euston and on the left Class 87, 87031 which entered traffic in July 1974. *(Bernard Mills)*

Birmingham at night and Class 25 5233 is seen at the east end of New Street and on the old Midland side. New to Traffic in December 1963 and outshopped in two tone green. May 1971 would see her painted rail blue and her tops number of 25083 would be added in February 1974. A Saltley allocated engine for two years from 1964 25083's subsequent career would see allocations including Glasgow Eastfield, Inverness and closer to home at Bescot and Crewe. Withdrawn in 1984 the engine is now preserved at Brechin. *(Bernard Mills)*

Above: The western end of New Street and the former Midland side sees Class 45 'Peak' 45075 on a west country bound working. Built as D132 and entering traffic in December 1961 she would continue in service until January 1985. On the middle road sits an unidentified class 08 as station pilot. *(Arthur Turner)*

Right: Again the west end of New Street and an AL5, class 85 E3078 is seen. Doncaster built for service on the West Coast Main line and entering traffic in March 1963, she would become 85023 in June 1974 and would survive in traffic until April 1990. The blue BRUTE parcels trolleys which were such a useful place for trainspotters to sit and stow their bags can be seen on all the platforms. Introduced in 1964 BRUTE was British Railways Universal Trolley Equipment they were loaded with parcels for a particular destination and loaded directly onto the coach. *(Bernard Mills)*

Birmingham
Shopping Centre
1S61
E3078

Looking at the Wolverhampton line exit from New Street and the bridge carrying Navigation Street over the station throat where we saw the tram earlier. A Midland Red bus makes its way out of Birmingham. Does anyone still stock Guernsey tomatoes now? Are they still a thing? Ansells used to be brewed in Aston but the brewery was closed in 1981 with production moving to Burton upon Trent. E3111 was an AL6 built at Doncaster and entering traffic in June 1965. Subsequently under TOPS became a class 86 and renumbered to 86024. Withdrawn as 86424 initially for preservation, she was exported to Hungary in August 2020. *(Bernard Mills)*

Birmingham West Suburban Railway post steam era and a three car DMU departing from Five Ways Station. Opened in 1885, it closed in 1944 as a result of competition with local bus services. It reopened in 1978 with the creation of the cross city line and introduction of through services from Lichfield to Redditch and Bromsgrove. In the background was the line into central goods station as seen on page 35. *(John Tolson)*

EMU 304045 arrives at Bescot Station on the 23 April 1991 with a Walsall bound service. Introduced in the early 1960s these units were initially four car and designated AM4 becoming Class 304 under the TOPS numbering system. On introduction they were the only AC electric units to carry BR Green livery. Refurbishment in the mid 1980's saw the sets reduced to three car units. 304045 was the last of the class built. The last Class 304s units survived until March 1996. To the right was the hump marshalling yard and the roof of the yard signal box can just be seen behind the catenary stanchion on the right. To the left are stabled locomotives of classes 20, 58 and 31. *(Andrew Royle)*

The other end of Bescot station and 310106 approaches with a special working. The lines curving away to the right are to Walsall, Hednesford and Rugeley, with those straight on to Wolverhampton. The impact of the M6 can be seen with the junction for Wednesbury on signposts and behind the photographer the M5 junction. The old Bescot engine shed can just be seen far left and survived in increasing dereliction to December 2013 when it was demolished. *(Andrew Royle)*

With the end of steam came the closure of Saltley shed and demolition of the roundhouses. It remained though a stabling point for diesels and a new three road shed was built, seen on the left. Saltley men become known as 'Seagulls' because their duties took them coast to coast and almost everywhere and they had a reputation for always keeping the job going. A fine collection of Class 47s. 45s, 25s and 20s are stabled on Sunday 25 November 1979. *(Arthur Turner)*

Two class 37s, 37688 and 37687 both in Railfreight livery are captured, having emerged from Washwood Heath sidings with empties for Peak Forest on 14 November 1987. The main lines are the former Midland Railway Derby to Birmingham line and looming over them are the gas holders which were such a dominant backdrop up to 2015. The bridge carries the Stechford to Bescot line. 37688 was built as D6905 and into traffic in November 1963 and is still in main line service with LSL at Crewe in 2025. 37687 entered service as D6881 in October 1963 and is also still in main line service with Colas. *(Andrew Royle)*

Opening in 1976 Birmingham International station is situated alongside Birmingham Airport and the National Exhibition Centre. In this 1987 view 47540 is seen departing southwards. New to traffic in March 1964 as D1723, she would become 47132 under TOPS numbering before having Electric Train heating and air braking fitted in October 1974 and renumbered 47540. Withdrawn in 1998 the engine was initially preserved but succumbed to scrap in 2016. *(John Tolson)*

The eastern approaches to New Street are amongst the busiest in the UK with over 800 movements a day. Between February 1999 and December 2000 the junction was extensively remodelled, structures rebuilt, signalling and power upgraded with £36m spent on the work which was undertaken by WS Atkins and Railtrack. This undated scene captures the Class 323 EMUs which were among the last trains to enter service with British Rail prior to privatisation and replaced the earlier Class 304 and Class 310 units. Designed in Birmingham but built by Hunslet in Leeds, they entered service in 1994. Distinctive of sound and with rapid acceleration these units would be the mainstay of cross city operations until they were replaced by Class 730 units in September 2024. The Rotunda once again dominates the scene and just below to the left can be seen the roof of Moor Street station that the New Street lines dive under. *(Milepost)*

Class 220 Voyager 221144 at Birmingham New Street in June 2009. These units have become the backbone of cross country services through Birmingham. Built by Bombardier in Belgium and introduced in 2001 they took over from HSTs and loco hauled workings with class 47s (and previously class 45s). Above can be seen the Pallasades Shopping Centre, predecessor to Grand Central. *(Andrew Royle)*

The last railway rolling stock assembled in Birmingham were the Class 390 Pendolino units by Alstom, who were successors to Metropolitan Cammell at their Washwood Heath Works. Constructed for the West Coast Main Line, fifty three units were ordered. Designed to tilt at speed these units were a development of British Rail APT technology by Alstom. Production at Washwood Heath ceased in 2005 and the works were demolished in 2019 but the site is to become a maintenance depot for HS2. *(Milepost)*

Left Top: A little oddity at Tyseley was the Allen Rowland scrapyard which could be glimpsed behind the goods loops. Occasionally a shunter could be seen. The business was as merchants of ferrous scrap for steelworks and ironfounders and wagon loads used to be despatched for Stocksbridge engineering steels at Deepcar, Sheffield. Motive power seen here were No. 6 and 7 which were John Fowler Diesel Mechanical 0-4-0s. The yard closed some time ago and for a long time the site was derelict; however a builders merchant now occupies the site. *(John Tolson)*

Left Bottom: When the line through Snow Hill was reopened in 1987 a new station was built at Moor Street on the main lines. This was fairly austere and was rebuilt again as part of the subsequent reopening of the original Moor Street Terminus platforms in a sympathetic GWR style. Very conveniently placed for the central shopping area of Birmingham the station now handles around seven million passengers per annum and nearly 1300 scheduled services a week. On the right is class 150 sprinter 150010. The class, introduced in 1985 would replace the first generation DMUs on services in the West Midlands and would be the mainstay of services until replacement by Class 172s in the late 2010s. On the left is a Chiltern 'Clubman' Class 168, 168005. Introduced in 1997 and built at Derby by Bombardier, these were the first stock ordered post the privatisation of British Rail. 24 June 2010. *(Andrew Royle)*

Right: December 2010 saw the old terminus platforms which had closed in 1987 with the reopening of Snow Hill brought back into use for the very reason Moor Street was originally built: lack of capacity at Snow Hill to cope with traffic. Chiltern Railways invested significantly to double the singled section of the former GWR main line to London, lift overall speeds and create an alternative Birmingham to London route. Vintage Trains marked the occasion with a Moor Street to London working, the ECS legs being piloted by GWR Pannier Tank 9600 which spent the day on display at the station. Here we see your author and his then very young eldest capturing the scene. *(Alistair Grieve)*

OUT
WAY OUT
WAY OUT
9600
35470

A look at Birmingham cannot pass without another look at Tyseley, now home to Vintage Trains and Tyseley Locomotive Works. 46233 *Duchess of Sutherland*, 5043 *Earl of Mount Edgcumbe* and 9600 are seen raising steam at an Open day in June 2017. *(Martin Creese)*

In the introduction we saw an image of New Street Station and Queens Hotel from Stephenson Street. This is the up to date view from the other end as an Edgbaston bound tram pulls away from its stop outside Grand Central on the right. The mirrored building is where Queens Hotel once stood. Navigation Street heads away to the right behind the photographer where we saw the original Birmingham Tram on page 11. The brick built building above the tram is the former Midland Hotel. *(Martin Creese)*

It seems fitting to finish where we started at Curzon Street and currently sat in the middle of a huge building site. The original Grand Central and London and Birmingham Railway station will have a new life as part of the HS2 Birmingham station whose elevated structure can be seen advancing.
The concourse will be almost next door to Moor Street. Tram tracks are also being extended to serve the new station and will loop under the elevated structure before heading to Digbeth. *(Martin Creese)*